KESHAV HARI S
Shajitha Banu S

ARTIFICIAL DEMAND IN LOGISTICSA SUPPLY CHAIN DISRUPTION

KESHAV HARI S
Shajitha Banu S

ARTIFICIAL DEMAND IN LOGISTICSA SUPPLY CHAIN DISRUPTION

An Empirical Study in Logistics and Supply Chain Disruption

LAP LAMBERT Academic Publishing

Cover image: www.ingimage.com

Publisher:
LAP LAMBERT Academic Publishing
is a trademark of
Dodo Books Indian Ocean Ltd. and OmniScriptum S.R.L publishing group

120 High Road, East Finchley, London, N2 9ED, United Kingdom
Str. Armeneasca 28/1, office 1, Chisinau MD-2012, Republic of Moldova, Europe
Managing Directors: Ieva Konstantinova, Victoria Ursu
info@omniscriptum.com

Printed at: see last page
ISBN: 978-620-8-44124-1

ACKNOWLEDGEMENT

Each and every work needs support and guidance for successful achievement of its aim and objectives. This dissertation works has also received support and guidance from many hands and above all the blessings of the Almighty God.

I am deeply grateful to the Managing Trustee of Sri Ramakrishna college of Arts and Science, Coimbatore, for providing me with this opportunity.

I am extremely thankful to **Dr. B.L. SHIVAKUMAR,** Principal & Secretary of Sri Ramakrishna college of Arts and Science, Coimbatore for having given me an opportunity to do this study.

I feel indebted to **Dr. I. PARVIN BANU M.com., M.Phil., MBA., M.com(CA), PGDCA, NET., Ph.D** Head of the PG & Research Department of International Business, Sri Ramakrishna college of Arts and Science, Coimbatore for having given me valuable support to complete this study successfully.

I would like to thank my guide **Mrs.S.SHAJITHA BANU M.Com., M.Phil., (Ph.D)** Assistant Professor in Commerce, and other faculty members PG and Research Department of International Business who have helped me towards the completion of this dissertation successfully.

I extend my sincere thanks to Librarian, Sri Ramakrishna college of Arts and Science for a timely help to finish my dissertation successfully.

I would also like to acknowledge my family members and friends who have been a source of courage and inspiration in all endeavours to complete this piece of work in time.

KESHAV HARI S

TABLE OF CONTENTS

S.NO	TITLE	PAGE NO.
1	CHAPTER – I	
1.1	Introduction	1
1.2	Statement of problem	7
1.3	Objectives of the study	7
1.4	Scope of the study	7
1.5	Research Methodology	9
1.6	Limitations of the study	10
1.7	Chapter scheme	10
	CHAPTER – II	
2	Review of literature	11
	CHAPTER – III	
3	Profile of the Company	21
	CHAPTER – IV	
4	Data Analysis and interpretation	25
	CHAPTER – V	
5	Findings, suggestions, and conclusion	45
6	Bibliography	47
7	Questionnaire	53

LIST OF TABLES

TABLE NO	CONTENT OF TABLES	PAGE NO
4.1	Age of the Respondents	27
4.2	Gender of the Respondents	28
4.3	What is your role in the supply chain/logistics industry	29
4.4	How many years of experience do you have in logistics/supply chain management	30
4.5	What is the size of your company's supply chain operations	31
4.6	Have you encountered artificial demand in your supply chain operations	32
4.7	What do you think are the key factors contributing to artificial demand in logistics	33
4.8	How often do you experience supply chain disruption due to artificial demand	34
4.9	What are the primary impacts of artificial demand in your logistics operations	35
4.10	To what extend has artificial demand affected your company's profitability	36
4.11	What are the key drivers of artificial demand in logistics	37
4.12	Have you noticed an increase in artificial demand due to global events (eg., pandemic, trade restrictions)	38
4.13	Do external factors like digital marketing and media exaggerate artificial demand	39
4.14	What measures does the company take to mitigate the effects of artificial demand	40
4.15	How effective are your current strategies in managing artificial demand	41
4.16	Do you think technology (AI, blockchain, predictive analysis) can help reduce artificial demand related disruption	42
4.17	Do you believe government regulations should be enforced to control artificial demand	43
4.18	What role do you think customers play in creating or mitigating artificial demand	44

LIST OF CHARTS

CHART NO	CONTENT OF CHART	PAGE NO
4.1	Age of the Respondents	27
4.2	Gender of the Respondents	28
4.3	What is your role in the supply chain/logistics industry	29
4.4	How many years of experience do you have in logistics/supply chain management	30
4.5	What is the size of your company's supply chain operations	31
4.6	Have you encountered artificial demand in your supply chain operations	32
4.7	What do you think are the key factors contributing to artificial demand in logistics	33
4.8	How often do you experience supply chain disruption due to artificial demand	34
4.9	What are the primary impacts of artificial demand in your logistics operations	35
4.10	To what extend has artificial demand affected your company's profitability	36
4.11	What are the key drivers of artificial demand in logistics	37
4.12	Have you noticed an increase in artificial demand due to global events (eg., pandemic, trade restrictions)	38
4.13	Do external factors like digital marketing and media exaggerate artificial demand	39
4.14	What measures does the company take to mitigate the effects of artificial demand	40
4.15	How effective are your current strategies in managing artificial demand	41
4.16	Do you think technology (AI, blockchain, predictive analysis) can help reduce artificial demand related disruption	42
4.17	Do you believe government regulations should be enforced to control artificial demand	43
4.18	What role do you think customers play in creating or mitigating artificial demand	44

CHAPTER 1
INTRODUCTION

1.1 INTRODUCTION

Artificial demand or manufactured demand constitutes demand for something that, in the absence of exposure to the vehicle of creating demand, would not exist. It has controversial applications in microeconomics (pump and dump strategy) and advertising.

A demand is usually seen as artificial when it increases consumer utility very inefficiently; for example, a physician prescribing unnecessary surgeries would create artificial demand. Government spending with the primary purpose of providing jobs (rather than delivering any other end product) has been labelled "artificial demand".Similarly Noam Chomsky has suggested that unchecked militarism is a type of government-created artificial demand, a "system of state planning ... oriented toward military production, in effect, the production of high technology waste", with military Keynesianism or a powerful military industrial complex amounts to the "creation of state-guaranteed markets for high technology waste (armaments)."

Artificial demand in logistics refers to the intentional creation or manipulation of demand for goods and services beyond their natural market levels. This can be driven by various factors, including marketing strategies, supply chain adjustments, government policies, or speculative market behavior. Unlike organic demand, which arises naturally from consumer needs and preferences, artificial demand is often stimulated by external interventions such as discounts, panic buying, production constraints, or strategic stockpiling.

In the logistics sector, artificial demand can have significant impacts on supply chains, inventory management, and transportation systems. For example, during peak shopping seasons or promotional events, companies may create artificial demand through limited-time offers or exclusive product releases, leading to increased shipping volumes and logistical challenges. Similarly, disruptions like sudden regulatory changes or trade restrictions can artificially inflate demand for certain goods as businesses rush to secure supplies.

Understanding artificial demand is crucial for logistics professionals, as it affects capacity planning, warehousing, and distribution efficiency. Companies must balance their response to short-term demand spikes while maintaining long-term supply chain stability. By leveraging data analytics, demand forecasting, and strategic partnerships, businesses can mitigate risks and optimize operations in the face of artificially induced market fluctuations.

1. Definition and Overview

Artificial demand in logistics refers to a demand that is not purely driven by genuine consumer needs but rather by external interventions such as marketing campaigns, supply chain strategies, or economic conditions. Unlike organic demand, which follows predictable trends based on consumer behavior and purchasing power, artificial demand can cause sudden spikes or declines in logistics operations.

2. Causes of Artificial Demand in Logistics

- Companies create artificial demand by offering limited-time discounts, flash sales, or exclusive product launches, leading to short-term surges in shipping and warehousing needs.

- Natural disasters, trade restrictions, or geopolitical events can trigger panic buying or hoarding, artificially inflating demand for certain products.

- Policies such as subsidies, tariffs, or export bans can influence demand artificially by making products more or less accessible in certain markets.

- Businesses or consumers may purchase large quantities of goods due to fear of shortages or anticipated price increases, leading to temporary supply chain strain.

- Product releases (such as new smartphones) or seasonal events (such as holiday shopping) can create artificial demand spikes that require careful logistical planning.

3. Impact on Logistics Operations

- Sudden demand fluctuations can lead to stockouts or overstock situations, affecting warehouse efficiency and increasing storage costs.

• If artificial demand is not accurately forecasted, suppliers and distributors may struggle to keep up, leading to delivery delays and bottlenecks.

• Logistics networks must handle increased shipping volumes, requiring additional resources such as trucks, shipping containers, and labor.

• Meeting artificial demand spikes may require expedited shipping, additional warehousing, or flexible distribution strategies, increasing operational expenses.

• Excess production and transportation due to artificial demand contribute to higher carbon emissions and resource consumption.

4. Managing Artificial Demand in Logistics

• Advanced analytics, AI, and historical data can help predict artificial demand trends and optimize inventory planning.

• Businesses can adopt just-in-time (JIT) inventory systems, diversify suppliers, or implement contingency plans to respond to fluctuations.

• Strong coordination between manufacturers, distributors, and retailers helps mitigate supply chain disruptions.

• Companies must develop contingency plans to address potential demand surges without overcommitting resources.

• Optimizing routes, reducing waste, and using eco-friendly packaging can help offset the negative environmental effects of artificial demand.

Understanding Artificial Demand in Logistics

Artificial demand refers to demand that is artificially created or influenced by external factors rather than natural consumer needs. This phenomenon significantly impacts logistics by causing unexpected fluctuations in supply chain activities, requiring businesses to adapt quickly to changes in demand patterns. Unlike organic demand, artificial demand often results in short-term spikes that may not be sustainable, leading to inefficiencies in inventory management, transportation, and distribution networks.

The Impact of Artificial Demand on Logistics Operations

A. Supply Chain Disruptions

• Artificial demand can cause overstocking or stockouts, leading to increased holding costs or lost sales opportunities.

• Minor fluctuations in consumer demand can be amplified throughout the supply chain, causing inefficiencies in production and distribution.

B. Logistics and Transportation Challenges

• Sudden demand spikes lead to higher transportation costs, as businesses rush to fulfil orders using express delivery options.

• Short-term demand surges require additional storage space, increasing operational expenses.

• E-commerce-driven artificial demand creates delivery bottlenecks, particularly in urban areas.

C. Environmental and Sustainability Concerns

• Artificial demand may lead to unnecessary production, increasing waste and environmental degradation.

• Increased transportation and expedited shipping contribute to a larger carbon footprint.

4. Strategies for Managing Artificial Demand in Logistics

• Leveraging big data and machine learning can help companies predict and respond to artificial demand fluctuations.

• IoT and AI-driven insights enable businesses to adjust logistics strategies dynamically.

• Implementing just-in-time (JIT) and dynamic inventory allocation can help prevent overstocking or understocking.

• Working with multiple suppliers reduces the risk of supply chain bottlenecks caused by artificial demand spikes.

• Businesses can adopt on-demand warehousing services to manage seasonal fluctuations without long-term storage costs.

• Companies can enhance delivery efficiency by leveraging route optimization technologies and alternative transportation modes.

• Reducing packaging waste and using electric or hybrid delivery vehicles can mitigate environmental impact.

• Encouraging recycling, reusing, and repurposing products can help counteract the negative effects of artificial demand.

Case Studies of Artificial Demand in Logistics

Panic Buying During the COVID-19 Pandemic

• In 2020, global supply chains were heavily disrupted due to artificial demand for essential items like hand sanitizers, toilet paper, and medical supplies.

• Logistics networks struggled to meet the surge in demand, causing shortages and price hikes.

• Companies responded by increasing local sourcing, implementing purchase limits, and using data-driven demand forecasting.

The Future of Artificial Demand in Logistics

• Predictive analytics will play a critical role in managing artificial demand fluctuations.

• Enhanced tracking and authentication can reduce artificial scarcity and speculation-driven demand spikes.

• Companies will prioritize green logistics strategies to mitigate the environmental impact of demand-driven inefficiencies.

• Shared data and real-time coordination will improve responsiveness to artificial demand changes.

Artificial demand plays a significant role in logistics, impacting supply chains, inventory management, transportation, and warehousing. As markets become more dynamic due to globalization, technological advancements, and changing consumer behaviours, understanding artificial demand becomes crucial for businesses aiming to maintain efficiency and sustainability.

Artificial demand plays a significant role in logistics, impacting supply chains, inventory management, transportation, and warehousing. As markets become more dynamic due to globalization, technological advancements, and changing consumer behaviours, understanding artificial demand becomes crucial for businesses aiming to maintain efficiency and sustainability.

1. Artificial Demand in Logistics

Artificial demand refers to the temporary or externally stimulated increase in demand for products and services, rather than natural consumer-driven demand. This demand can be influenced by marketing tactics, policy changes, speculative buying, or sudden economic shifts.

Unlike organic demand, which is based on long-term consumer needs and purchasing behaviour, artificial demand often creates short-term spikes that can lead to supply chain disruptions, increased costs, and inefficiencies in logistics operations.

2. Categories and Causes of Artificial Demand in Logistics

Artificial demand can be broadly classified into planned artificial demand and reactive artificial demand:

A. Planned Artificial Demand

This type of demand is intentionally created by businesses or institutions to drive sales, enhance brand positioning, or influence purchasing behaviour.

1. **Marketing and Promotional Strategies**
   - Flash sales, seasonal discounts, and exclusive product launches (e.g., Black Friday, Cyber Monday).
   - Limited-time offers that create urgency and encourage bulk purchasing.

- Brand partnerships with influencers to create hype around products.

2. **Strategic Supply Chain Constraints**
    - Companies deliberately restricting product supply to create perceived scarcity.

1.2 STATEMENT OF THE PROBLEM

- Many small scale logistics companies have no knowledge about it .
- It directly affects the GDP "Gross Domestic Product".
- Many consumers are influenced and affected financially and they are not more often adaptable about the stock availability in order region.
- Many third part agents are selling the products with a higher rate without a limit.
- The lack of comprehensive research and understanding of artificial demand in the logistics sector further exacerbates the problem.

1.3 OBJECTIVES OF THE PROBLEM

- To analyse the stock availability with the help of tracking the demand products by using a server which can track the products.
- To avoid the purchase of foreign products without knowing the actual price.
- To purchase products after knowing it clearly so it can control the high price demand.
- To make a long term relationship with the known third party agents.
- To propose strategies and frameworks for mitigating artificial demand in logistics.

1.4 SCOPE OF THE STUDY

The study on artificial demand in logistics and its impact on supply chain disruptions focuses on understanding the factors contributing to the creation of artificial demand and its subsequent effects on supply chain operations. The scope of this research includes the following dimensions:

Conceptual Scope:

Demand created by marketing (e.g., influencer marketing, psychological pricing). Demand due to planned obsolescence (e.g., tech industry, fast fashion). Supply-driven demand manipulations (e.g., limited product release, artificial scarcity). Regulatory or policy-driven artificial demand (e.g., government subsidies, tax incentives).

Industry-Specific Scope:

The study focuses on artificial demand in logistics prevalent to:

Technology and Electronics

Planned Obsolescence : Examining how companies design products with a limited lifespan to drive continuous demand.

Scarcity Marketing: Impact of limited-edition product launches (e.g., iPhones, gaming consoles.)

Fashion and Luxury Goods

Fast Fashion vs. Luxury Branding: How artificial demand is created through trends and exclusivity.

Hype Culture: The role of limited-edition releases and celebrity endorsements.

Sustainability Concerns: Ethical implications of forced consumption cycles.

Pharmaceuticals and Healthcare

Drug Demand Manipulation: Exploring the role of aggressive marketing in over prescription.

Vaccine Hoarding and Supply

Control: Impact of artificial scarcity during health crises.

Regulatory Influence: Examining government policies that drive artificial demand for medications.

E-commerce and Retail

Flash Sales and Limited-Time Offers: The psychological impact of urgency marketing.

Black Friday and Holiday Shopping Events: How artificial demand spikes affect consumer spending.

Subscription Models: Examining how companies lock consumers into continuous demand cycles.

Industry Scope: The research will examine multiple industries, including retail, healthcare, e-commerce, manufacturing, and consumer goods, where artificial demand is frequently observed during disruptions. Special attention will be given to industries with high sensitivity to supply-demand imbalances.

Temporal Scope: The study will focus on both historical and contemporary supply chain disruptions to identify trends and recurring patterns in artificial demand. Events such as the COVID-19 pandemic, natural disasters, and geopolitical conflicts will serve as case studies to provide context.

1.5 RESEARCH METHODOLOGY

Research Approach

This study adopts a mixed-methods research design, incorporating both qualitative and quantitative approaches to analyze artificial demand and its impact on supply chain disruptions.

• Exploratory Approach: To understand the concept of artificial demand and its role in supply chain disruptions.

• Descriptive Approach: To examine real-world case studies and trends in artificial demand-driven disruptions.

• Causal Approach: To identify cause-and-effect relationships between artificial demand and supply chain inefficiencies.

Data Collection Methods

A. Primary Data Collection

Collected through google forms.

B. Secondary Data Collection

Collected from the company websites and resources.

Tools Used

Percentage analysis

1.6 LIMITATIONS OF THE STUDY

- The respondents are limited to 50.
- The study is limited over the specific area of the company.
- No confidential informations were shared from the company.

1.7 CHAPTER SCHEME

The present study is separated into five divisions, each with its own set of tables and appendices in order to accomplish the aforementioned objectives.

- CHAPTER 1- The first chapter covers introduction, statement of the problem, objectives, scope of the study , Research methodology, limitation of the study and chapter scheme.

- CHAPTER 2 - The second chapter gives covers the Review of Literature.

- CHAPTER 3 - The Third chapter discusses Overview of the Study.

- CHAPTER 4 - In chapter fourth it includes, the data analysis and interpretation were presented methodically, with tables and displays.

- CHAPTER 5 - This chapter summaries the findings and suggestions. The conclusion is drawn using the study's findings.

CHAPTER 2

REVIEW OF LITERATURE

In 2003, Pienaar traces the origin of logistics, in military usage and its subsequent extension to commercial usage. The author suggests that the definition of logistics management given by the Council of Logistics Management (CLM) can be adjusted to read as follows: "Logistics Management is the process of planning, organizing and executing the efficient effective flow and storage of goods, services and related information from the place of origin to the place of consumption or application for the purpose of optimally meeting customer and client requirements in order to help maximize the long-run welfare of the firm". The important determinants of logistics service performance are:-- suitability, accessibility, goods security, transit time, reliability, and flexibility.

In 2006, Balloudeals with the past, present, and future of logistics and supply chain management. The author traces the major events that created business logistics as it is practiced today, attempts to summarize the present state of business logistics and how it relates to supply chain management (SCM). As regards the future, some predictions are made as to the issues that need to be addressed and the events that will likely take place in the near term. Logistics is now viewed as a subset of SCM.

In 2006, Guarnieriet. al., suggest that trough the constitution of entrepreneurial agglomerations of suppliers, it is possible to obtain logistic advantages in the transporting, stocking, and warehousing activities and maximize SCM competitiveness through the establishment of trust, and lasting relationships between the components of the whole chain. The entrepreneurial agglomerations of suppliers results in a gain for all those participating in the chain.

In 2006, Sánchez, et. al., state that manufacturing firms are increasingly placing greater emphasis on the management of their outsourced distribution channels (DC). However, the role that inter-organizational Management Control Systems (MCS) can play in managing DC problems is still not clearly understood. MCS can help to complement and re-orientate inter-firm agreements and constitutes a key tool for managing DC's in a flexible way. Vertically integrating the distribution function is not the only option for the manufacturer to gain control.

In 2007, Barutcu highlights business, marketing and manufacturing strategies so as to supply customized products efficiently. Many of the customers are ready to pay premium price for customized products. Hence companies are forced to respond to growing individualization of demand, have the ability to change the product to meet individual needs, and finally respond to customized products. Supplying customized products program can succeed if a company integrates relationship marketing, mass customization, agile manufacturing, SCM, efficient logistics operations, and information systems.

In 2007, Borade and Bansod attempt to provide an understanding of the supply chain activities and its mapping by answering the rationale behind SCM. They discuss the concept of SCM along with its need from organization point of view. There are interrelations and dependency among business operations. Current trends like outsourcing, information technology adoption, and third party logistics present an opportunity for development of SCM. Organizations will have to adopt partnership information sharing initiative with suppliers and establish mutual trust within the supply chain for effective SCM practice.

In 2007, Chieh-Yu and Yi-Hui discuss about how China's Logistics Industry has started to pay attention to adopt more efficient logistics technologies to provide better services to customers. This will increase supply chain performance for the logistics industry in China. Most of China's Logistics service providers place emphasis on the innovation in information technologies to enhance their supply chain performance.

In 2008, Rahman, et. al. discuss about current supply chain practices and its barrier in order to help Malaysian manufacturing companies to manage and solve their supply chain problems. The authors conclude that there are interesting barriers between companies in Malaysia and other countries in respect of SCM implementation. They suggest that problems can be solved by focusing on SCM development and exposing the employees to SCM implementation.

In 2008, Kara, et.al discusses about challenges in managing variety of risks during the complete product design, development and delivery process. The relationships that develop between a purchasing company and its suppliers as well customers result in networks or supply chains that are more exposed and vulnerable to detrimental events. The authors develop a framework to categorize the sources of risks into three main areas:-- environment, supply chain, and organization, with unique risk classifications. This can be used as a risk identification tool and also to establish inter- and intra-relationships between various risk factors.

In 2009, Ezutah and Kuan identify effective supply chain performance measurement as a key issue towards efficient SCM. The authors examine both traditional and fuzzy logic approaches to supply chain performance measurement. Based on the study, the authors highlight the benefits accruable from the use of fuzzy logic operation in supply chain performance measurement. Finally, they recommend on applying fuzzy logic operations in measuring the performance of a green or sustainable supply chain.

In 2009, Shao, et. al. design a systematic and comprehensive hybrid index evaluation system of regional logistics competitiveness according to the characteristics of China's regional logistics system under supply chain globalization environments. Overall, the regional logistics industry in China has made gratifying development in recent years, but the development level of logistics industry in the provincial capital cities in China remain seriously unbalanced. The regional economic development is increasingly dependent on the development level of the regional logistics industry system.

In 2010, Awad and Nassar discuss about extension of supply chain from upstream and downstream to include other vendors. To succeed in the digital economy, organizations must manage the integration of business, technology, people, and processes, not only within the enterprise but also across extended enterprises.

In 2010, Danuta introduces "Virtual Supply Chain" as an emerging concept in current business reality. The attributes of virtual supply chains are-- temporary character, focus on customers, geographical dispersion, intensive support of I.T. systems, network structure, and extensive use of key competencies of their members. Based on the research, the author concludes that the level of logistics and supply chain qualifications has been changing constantly. There is growing "Logistics Awareness" of operational and managerial staff in supply chain and logistics.

In 2010, Konecka discusses about Lean and Agile Supply Chain Management, widely discussed in literature. However, there are some disputes concerning differentiating these types of managements from tools and instruments such as outsourcing, single-sourcing, quick response, standardizing principles and postponement. "Supply Chain Risk Management" (SCRM), emphasizes the importance of risk management in supply chains, to achieve competitive advantage of those chains in which a risk is identified, estimated, managed and controlled. The concepts of lean and agile SCM could be viewed as concerning supply chain risk management and could facilitate the choice of a supply chain strategy based on risk analysis. The SCM concepts presented by the author have strong relationships with trends such as importance of reducing stock levels, mitigation of risk of economic slowdown, shortening the supply chain, increasing the visibility of supply chain etc.

In 2010, Korrakot, et. al discuss about SCM playing a major role in building a sustainable competitive edge for products in highly competitive markets. The Logistics Score card (LSC) is used as data collection tool. The authors conclude that high and low technology companies may develop different practices to achieve what they aim for.

In 2010, Olugu, et. Al bring out the ecological issues arising from manufacturing operations which led to a focus on environmental sustainability in manufacturing. In this study the authors propose a Closed Loop Supply Chain (CLSC) performance measurement in the automotive industry. The authors recommend that this approach be used against the traditional supply chain method. Finally they recommend that suitable metrics be developed for the assessment of CLSC performance.

In 2010, Pereira and Geiger carry out analysis of case studies of large, small and medium enterprises. The main challenges faced by SMEs are:-- the differences in organizational objectives and the little know-how about the reality of B2B transactions in this sector. Large companies offer non-profitable items to SMEs who are not active in the chain. They also do not get financial support from the large companies. Other challenges are high performance demands, prices imposed by big buyers, and lack of technical and managerial expertise of SMEs.

In 2010, Verma and Seth bring out how competitive priorities have forced organizations to change dramatically due to rising customer expectations, continually increasing competition on a world-wide scale, time and quality based competition, and mass customization. Modern competition has already become the competition among enterprises and their supply chains rather than the competition among enterprises only. Supply chain Competitiveness (SCC) can be obtained by the integrated efforts of all the components of the supply chain.

In 2011, Geethan et. al, explain the linkages between causes and effects through a case study. The performance metrics are calculated in terms of Return Value, Gatekeeping Effectiveness, Warehousing Effectiveness, Carrying Cost percentage, Recovery Efficiency, Recovery Rate, Environmental Effectiveness, Overall Vehicle Effectiveness, and Return Transit Time. The authors conclude that though the implementation of reverse logistics may be a risky endeavor as it involves financial and operational aspects, but, however, with legislative measures tightening up, there are not many options. The question is not whether to go for it, but which framework to pick up.

In 2011, Golinska and Kawa provide a framework for management of reverse flow of materials in automotive industry. They review the remanufacturing and traditional manufacturing. Simulation model is used. The uncertainty caused by "mismatch between supply and demand with respect to timing and quantity in recovery networks" is the main problem in materials management of combined forward and reverse flows. They suggest application of agent technology to help solve problems that appear very often in recovery networks.

In 2011, Jie-Shin and Jerry discuss the evolution process of Taiwan's economic and industrial development which has been profoundly affected by global and regional economies. With the changing role of China, Taiwanese enterprises are faced with opportunities and threats far greater than any other countries because of its geographical proximity and shared culture and language. When Taiwanese companies transitioned from original equipment manufacturers (OEM) to Original Design Manufacturing (ODM), it not only strengthened partnerships with international companies but also caused value-added business processes to upgrade from value-added process innovation to value-added product innovation. The authors propose four types of value-added models of logistics– manufacturing, distribution, reverse logistics, and integrated logistics.

In 2011, Madhusudhana discusses about how delivery performance provides an indication of how successful the supply chain is at providing products and services to the customer. It integrates the measurement of performance right from supplier end to customer end. The analysis by the authors helps the companies to bench mark their performance as well as the performances expected from their counter parts for successful SCM in terms of delivery performance.

In 2011, Morana and Seuring brings out how Supply chain management (SCM) and closed loop supply chain management (CLSCM) have developed into established concepts in recent years. The authors propose a framework for CLSCM spanning three different levels– the societal or governance, the chain and the actor level. The three level framework proposed by the authors integrated CLSCM and related business processes with their implications on the societal as well as the single actor level.

In 2011, Shekari, et. al discuss about how Green Supply Chain Management (GSCM) has emerged as a pro-active approach for improving environmental performance of processes and products in accordance with the requirements of environmental regulations. Thirty four measurements are developed on the basis of opinions from industrial experts. The results indicate that a six-factor measurement model including internal environmental management, green purchasing, cleaner production, recovery, eco-design, and pollution fits the data acceptably. GSCM has emerged as an important archetype for companies to achieve profit and market share efficiency.

In 2011, Shukla and Garg present a step-by-step approach for understanding a complete picture of supply chain. Current trends like outsourcing, information technology adoption, and third party logistics present an opportunity for development of SCM. In future, all organizations will have to adopt partnership information sharing initiative with suppliers. Establishing mutual trust within the supply chain will be great challenge for all.

In 2012, Bhanu Krishna et. Al focus on the importance of transportation and logistics in Green Supply Chain Management (GSCM). They explain their concepts with the help of four methodologies– Reverse Logistics, Calculating the Amount of CO2 released, shipment Consolidation, and Environmental Performance Index (EPI). Organizations should realize that GSCM can reduce the ecological impact of industrial activity without sacrificing quality, cost, reliability, performance or energy utilization, thus, leading to overall economic profit.

In 2012, Björnfot and Torjussen bring out that in the construction sector, large contractors generally collaborate across supply chains in matters like construction regulation, transportation taxes etc. but the plights of small and medium sized enterprises (SME) are many. According to the authors, horizontal supply chain collaboration among the SME is an important factor for survivability of SME. Horizontal supply chain collaboration provides a flexible business climate leading to improved SME competitiveness and survivability in a volatile market.

In 2012, Devi discusses about E-commerce. The volume of trade conducted electronically has grown dramatically since the spread of the internet. The private sector should take the lead role in the development and use of e-commerce, with the government playing a pivotal role by creating a favorable policy environment for e-commerce and providing e-government services to its citizens to encourage its mass use. At the same time, e-marketers need to address some of the issues face by on-line shoppers.

In 2012, Ilinca and Andreea present the key elements of performance measurement in relation to business growth, also taking into account the main challenges and obstacles. The authors bases their study on the research developed inside four companies from an extended supply chain in automotive industry, relating theoretical discoveries with real business experience. The lack of clear identification of performance factors, lack of integration of risk management and lack of integration of company's extended supply chain to senior suppliers brings enormous losses in times of economic crisis. There are two basic ways to evaluate an activity and increase its performance:-- counting and appreciation. Counting is generally recognized as the best measurement method because it is more objective. Appreciation is generally considered to be more subjective and ranks second.

In 2012, Jereb, et. alfocus on the risks by defining them by different key dimensions, so that risk management is simplified and can be undertaken in every supply chain and organizations within them. A freely accessible risk catalogue is assembled and available on line which can serve as a checklist. Every user of the model and the catalogue can find it as a new approach to supply chain risk management which is based on a detailed description of every identified risk.

In 2012, Kalenga and Brent present the South African automotive industry which faces increasing challenges to produce high quality products at low costs. The authors introduce a conceptual model whereby localization strategies of OEMs, with respect to their main suppliers can be optimized for competitive advantage. Based on the findings of a case study, the authors propose an improved model that could be used as a guideline to orientate procurement strategies for optimal performance. The model proposes orientation change strategy through five phases.

In 2012, Kayes and Azeem deal with incoming material management of a 100% export oriented a knit composite factory in Bangladesh. Different techniques of demand forecasting are implemented to find the best suitable model for a specific raw material. Regarding supplier selection, Analytical Hierarchy Process (AHP) technique is implemented to select the best supplier of the concerned raw material. The adaptive exponential smoothing method can forecast the future demand for a particular type of yarn selected in this study, very precisely. The AHP methodology is found to provide a better solution for selecting the best suitable supplier.

In 2012, Mohanti and Gahan focus on three aspects– Firstly to understand the buyer supplier relationship attributes considered by the Indian large scale manufacturing organizations, secondly expectations of large scale buyers from the MSME sector suppliers and finally the ranking of factors affecting the buyer supplier relationships. Buyer supplier relationship largely depends upon four major aspects. These are strategic requirements of the organization, supplier performance, mode of operation and personal factors.

In 2012, Xiao puts forward a definition of cluster supply chain as a new type of management pattern which is a combination of industrial cluster and supply chain management (SCM). The author gives a case study of collaborative procurement in a textile and garment industry cluster to demonstrate the construction and development of cluster supply chain, in which the role of service system is described in detail.

In 2013, Ben Rubenet. al explain the concept of Agile Supply Chain Management (ASCM) with the help of a case study in an automobile horn manufacturing company. In today's competitive scenario, organisations should be able to quickly respond to the volatile market. For this, the company has to become more agile and flexible. The authors suggest improvements for agility in the supply chain of the product.

In 2013, Craig presents a brief discussion of supply chains for flexible manufacturing systems with a particular focus on maintenance scheduling. Maintenance systems are created for ensuring the serviceability and safety of equipment or systems so that maximum possible performance levels can be achieved. Efficient maintenance scheduling is an important tool for any manufacturing environment.

In 2013, Koblen and Niznikova focus on explanation of main specifics of SCM in aerospace industry– OEM, processes and requirements for supplier selection,-- and subsequently the role and mission of international organisations involved in aerospace SCM and quality issues. For an organization, the understanding of its SCM requirements is the first key step. The next step is deciding whether to build a supply chain infrastructure in-house or partner with a supply chain product/service provider. The operational best practices can be utilized to implement a successful SCM solution in the framework of organization.

In 2013, Movahedi, et. al in their study, take Deming's principles as independent variables, organizational productivity is dependent variable and the study examines six different hypotheses. The authors present an empirical investigation on the effects of QMS implementation to organizational productivity through implementing change management. The model proposed in this study could contribute to productivity improvement.

In 2013, Nayak, bring out that as competition intensifies, companies move away from vertical integration of all operations to focus on core competencies, outsourcing other operations. However, with too many linkages, the supply chain becomes unresponsive to customer needs. This leads to development of value chain management (VCM). The author analyses the key aspects of VCM as integrated supply chain planning and scheduling, full resource management, cycle time responsiveness, chain-wide resource optimization and information integration. He also discusses the need for unleashing innovation across the Value Chain and brings out the benefits of systematic change in value chain from conventional to improvised value chain.

In 2013, Rakheja, et. al discuss about how in increasing competitive market, decision making plays an important role as the results depend upon the direction of concrete decisions taken. In this situation, theAnalytical Hierarchy Process (AHP) methodology proves to be very useful. The basic principles of AHP are:-- Principle of Decomposition, Principle of Comparative Judgments, and the principle of Hierarchic Composition/Synthesis. AHP will prove to be a better and more efficient source for the decision making process.

CHAPTER 3

OVERVIEW OF THE STUDY

CRYSTAL CARGO., Transport started its operations in Tirupur in 2009. However it was only in 2009 that CRYSTAL CARGO Transport was registered as a transport company. Since inception, CRYSTAL CARGO., Transport has been providing logistic services in the four District of Coimbatore, - Chennai Greater Coimbatore municipal area is the prime focus of our company for operations. However we also have the required skills, As a part of our service offering, we provide end to end logistics management as our flagship offering. It is a complete packaged solution that takes care of loading, unloading, needs, transportation and responsibility of the goods being transported.

- Domestic Transport
- Part Load
- Load
- Full Truck Load

Company Overview

Fastest Flourishing NVOCC

Inox shipping line, established in 2017 with the intent of serving the ever increasing trade. It's headquartered in Singapore, with offices in India Malaysia and Dubai.

Inox Shipping Line covers the entire India with 28 ICDs (In-land Container Depots), Southeast Asia, and the Middle East including China, Indonesia, Vietnam, Singapore, Malaysia, Jebel Ali, Shuwaikh, Bahrain, Umm Qasr, Oman, and further planning to add Red Sea Ports like Salalah, Djibouti and more. The success of Inox Shipping Line can be directly attributed to a skilled and experienced team that collaborates with clients to provide transport solutions that are customized to client's needs and are regularly examined to produce better outcomes each time. A tracking system that enables clients to track their freight at any time is completely supported by this. The firm has its own proprietary software that can be accessed through the system and handles all customer requirements directly through their own portal.

"We seek to innovate every day to be unique from other shipping lines in the market and gain a competitive edge".

This success can be directly attributed to a knowledgeable and experienced team who work with customers to tailor transport solutions that meet your requirements and are continually reviewed to deliver better results every time. This is fully supported with a Tracking system that allows you to trace your freight at any time. INOX holds a principal attitude towards its customers to provide best solutions in the aspect of reliability, honesty and competence. A personalized service is the key word of our specialized staff, assisting traders to move their goods both inland and overseas on a worldwide basis. Beside arrangements for routing, pricing and documentation the company also advises stowage matters and handles transport insurance.

Company History

INOX Shipping Line is a completely bootstrapped company that has grown exponentially.

The Growth Journey, INOX Shipping Line is a completely bootstrapped company that has grown exponentially since its inception in 2017 and created a prominent name itself in the industry. Over a short span of time, the firm has grown to a total turnover of 12 million dollars. The firm began operations in Navi Mumbai with 16 boxes and has since expanded to 3000 units. And as it forges ahead, the firm is driven by the mission of becoming the fastest NVOCC that covers the entire world. It is often seen that most NVOCC prefer doing business only in Southeast Asia, while INOX Shipping Line intends to expand into the US, Europe, and Africa.

Freight Management

It ensures that the cargo reaches its destination on time, in the right condition and cost-effectively.

Complete Logistics

Our international network allows us to handle shipments directly from terminal to terminal or even from door to door. We will find alternative transport solutions.

Dry Cargo

Our dry cargo service is the simplest and most common one we offer. It has been developed to ensure the safe and efficient shipment of any containerised cargo.For most of our customers, this service provides the support and equipment to move some of the world's most commonly traded commodities such as paper, white-goods and scrap metals: however we understand that not all cargoes are quite this straightforward.

Ports, Terminals and Depots

A global network of strategically-positioned ports enabling maximum customer choice and supply chain efficiency.

Hazardous Cargo Logistics

When we transport your hazardous cargo, we take the whole supply chain into consideration. From offering specialist training to our drivers, to the careful planning of container stowage on our vessels, we've established our operations to limit the risk associated with this type of cargo.

Cross Trading

At INOX we understand that sometimes our customers need to move cargo to and from countries in which they themselves aren't based. That's why we've established our local cross trades team, who can help with queries and bookings outside of the country from where your business operates. Our cross trades teams support our full range of containerised freight services, in particular dry cargo and hazardous cargo. Our teams provide a comprehensive service that includes competitive pricing, securing bookings, documentation handling, tracking facilities and pre/on carriage arrangements. This means that wherever your cargo is traveling to or from, you'll have a dedicated team supporting you, locally.

CHAPTER 4
DATA ANALYSIS AND INTERPRETATION

PERCENTAGE ANALYSIS

4.1 DATA ANALYSIS

Data analysis is the act of taking raw data and turning it into information that users can use o make decisions. In order to provide answers, test hypotheses, or refute ideas, data is gathered and examined by John Turkey, a statistician, in 1961 defined data analysis "Procedures for analysing da, techniques for interpreting the results of such procedures, ways of planning the gathering of data to make its analysis easier, more precise or more accurate, and all the machinery and results of (mathematical) statistics which apply to analysing data."

4.2 INTERPRETATION

Interpretation refers to the implementation of process through which data is reviewed of the purpose of arriving at an informed conclusion. The interpretation of data assigns a meaning to the information analysed and determines its significations and implications. The importance of a interpretation is evident and this is why it needs to be done properly. Data is very likely to give from multiple sources and has a tendency to enter the analysis with haphazard ordering.

4.3 FINANCIAL STATEMENT

A financial Statement is an organized collection of data according to logical and consistent accounting procedures. It purpose is to convey an understanding of some financial aspects of a business firms. It may show a position at a moment of time as in the case of a balance sheet, or Lay reveal a series of activities over a given period of time, as in the case of an income statement.

Thus, the term financial statement generally refers to the basic statement;

- The income statement
- The balance sheet
- A statement of retained earnings

TABLE SHOWING AGE OF THE RESPONDENTS
TABLE 4.1

S.NO	AGE GROUP	NO OF RESPONDENTS	PERCENT
1	18-25	35	70%
2	26-35	6	12%
3	36-45	6	12%
4	Above 45	3	6%
	TOTAL	**50**	**100**

INTERPRETATION:

Primary Data:

The table 4.1 implies that 70% of the respondents are from 18-25 age group, 12% of the respondents are in the age of 26-35, 12% of respondents are in the age of 36-45, 6% of respondents are Above 45.

Majority of the Respondents (70%) are belonging to Age Group of 18-25.

CHART 4.1

CHART SHOWING AGE OF THE RESPONDENTS

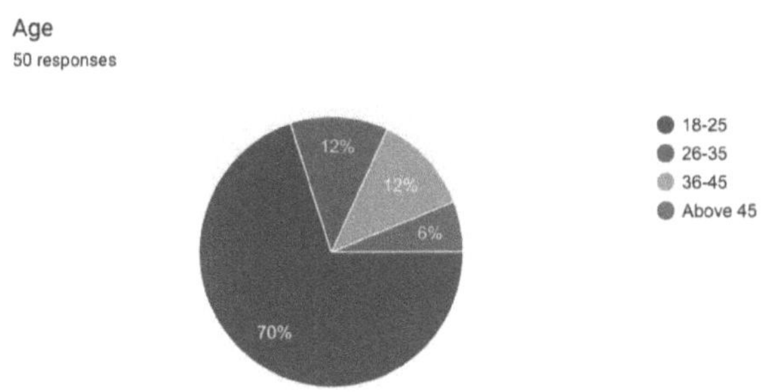

TABLE SHOWING GENDER OF THE RESPONDENTS

TABLE 4.2

S.NO	GENDER	NO OF RESPONDENTS	PERCENT
1	Male	35	71.4%
2	Female	14	28.6%
3	Others	0	0%
4	Not answered	1	0%
	TOTAL	**50**	**100**

INTERPRETATION:

Primary Data:

The table 4.2 implies that 28.6% Of Respondents are female, 71.4% of Respondents are male.

Majority of the Respondents (71.4%) are Male.

CHART 4.2

CHART SHOWING GENDER OF THE RESPONDENTS

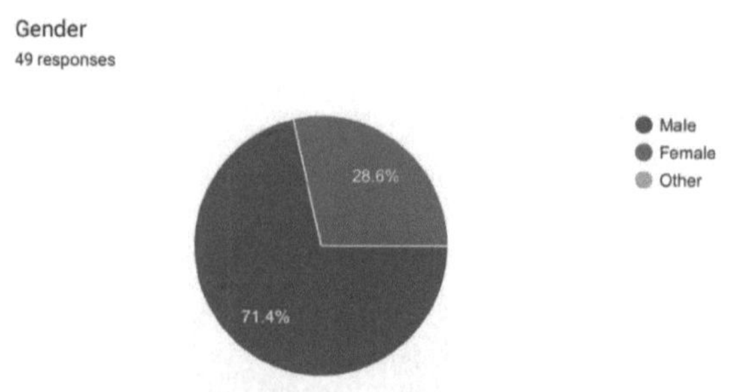

TABLE SHOWING THE ROLE OF THE SUPPLY CHAIN/LOGISTICS INDUSTRY

TABLE 4.3

S.NO	ROLE	NO OF RESPONDENTS	PERCENT
1	Manufacturer	11	22.4%
2	Supplier	14	28.6%
3	Distributor	13	26.5%
4	Retailer	5	10.2%
5	Others	6	12.2%
6	Not answered	1	0%
	TOTAL	**50**	**100**

INTERPRETATION:

Primary Data:

The table 4.3 implies that 22.4% of respondents are Manufacturer, 28.6% of respondents are Supplier, 26.5% of respondents are Distributors, 10.2% of respondents are Retailers and 12.2% of respondents are Others

Majority of the Respondents (28.6%) are Supplier.

CHART 4.3

CHART SHOWING THE ROLE OF THE SUPPLY CHAIN/LOGISTICS INDUSTRY

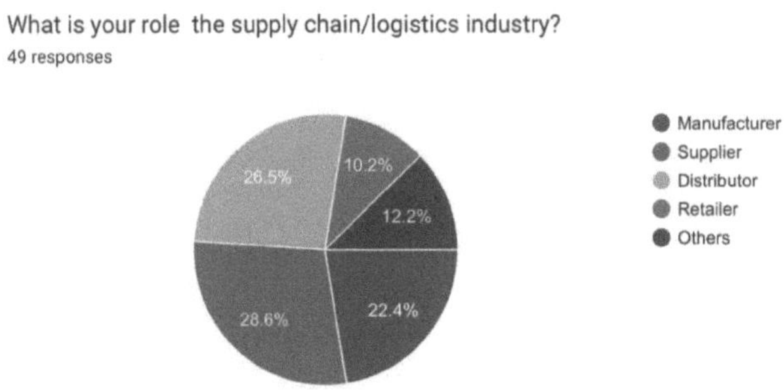

TABLE SHOWING THE YEARS OF EXPERIENCE IN LOGISTICS/SUPPLY CHAIN MANAGEMENT

TABLE 4.4

S.NO	EXPERIENCE	NO OF RESPONDENTS	PERCENT
1	Less than 1 year	21	42%
2	1-3 years	19	38%
3	4-7 years	6	12%
4	8+ years	4	8%
	TOTAL	**50**	**100**

INTERPRETATION:

Primary Data:

The table 4.4 implies that 42% of the have less than 1 year of experience , 38% of the respondents have 1-3 years of experience , 12% of respondents have 4-7 years of experience and 8% of respondents have experience of more than 8 years.

Majority of the Respondents (42%) have less than 1 year of experience .

CHART 4.4

CHART SHOWING THE YEARS OF EXPERIENCE IN LOGISTICS/SUPPLY CHAIN MANAGEMENT

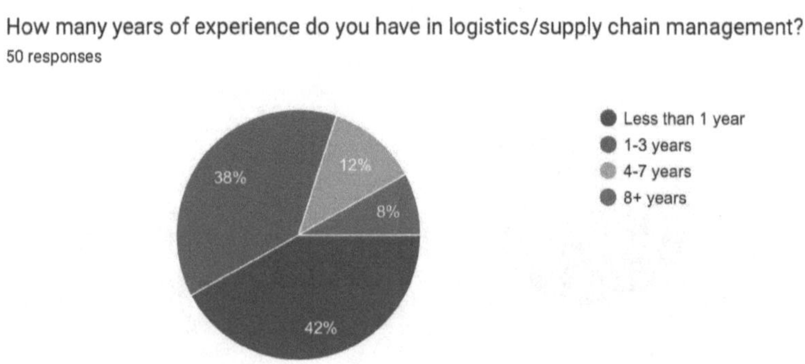

TABLE SHOWING THE SIZE OF COMPANY'S SUPPLY CHAIN OPERATIONS

TABLE 4.5

S.NO	COMPANY SIZE	NO OF RESPONDENTS	PERCENT
1	Small-scale	15	30.6%
2	Medium-scale	30	61.2%
3	Large-scale	4	8.2%
4	Not answered	1	0%
	TOTAL	**50**	**100**

INTERPRETATION:

Primary Data:

The table 4.5 implies that 30.6% of the respondents are from Small-scale company, 61.2% of the respondents are from Medium-scale company and 8.2% of respondents are from Large-scale company.

Majority of the Respondents (61.2%) are from Medium-scale company size.

CHART 4.5

CHART SHOWING THE SIZE OF COMPANY'S SUPPLY CHAIN OPERATIONS

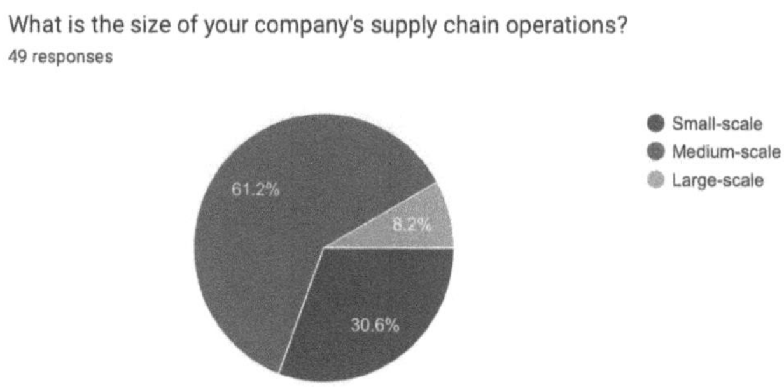

TABLE SHOWING ARTIFICIAL DEMAND IN SUPPLY CHAIN OPERATIONS
TABLE 4.6

S.NO	STATEMENT	NO OF RESPONDENTS	PERCENT
1	Yes	22	44%
2	No	15	30%
3	Maybe	13	26%
	TOTAL	**50**	**100**

INTERPRETATION:

Primary Data:

The table 4.6 implies that 44% of the respondents have stated YES, 30% of the respondents have stated NO and 26% of respondents have stated MAYBE.

Majority of the Respondents (44%) have stated YES.

CHART 4.6

CHART SHOWING ARTIFICIAL DEMAND IN SUPPLY CHAIN OPERATIONS

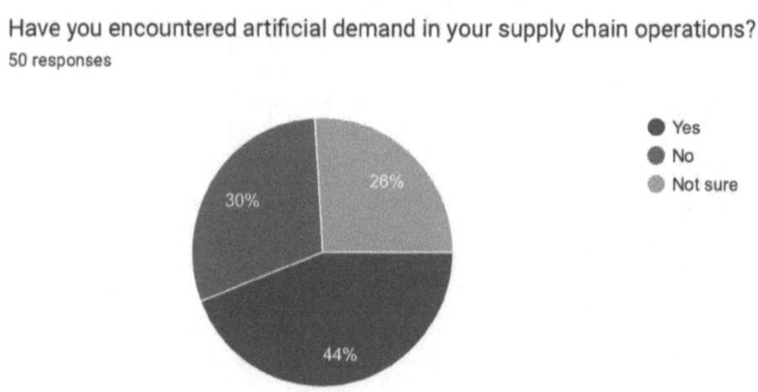

TABLE SHOWING THE KEY FACTORS CONTRIBUTING TO ARTIFICIAL DEMAND

TABLE 4.7

S.NO	KEY FACTORS	NO OF RESPONDENTS	PERCENT
1	Hoarding of inventory	7	14%
2	Speculative purchasing	15	30%
3	Supply chain manipulation	15	30%
4	Market speculation	10	20%
5	Others	3	6%
	TOTAL	**50**	**100**

INTERPRETATION:

Primary Data:

The table 4.7 implies that 14% of the respondents agree to Hoarding of inventory, 30% of the respondents agree to Speculative purchasing, 30% of respondents agree to Supply chain manipulation, 20% of respondents agree to Market speculation and 6% have voted for Others

Majority of the Respondents (30%) have voted for Speculative purchasing and Supply chain manipulation .

CHART 4.7

CHART SHOWING THE KEY FACTORS CONTRIBUTING TO ARTIFICIAL DEMAND

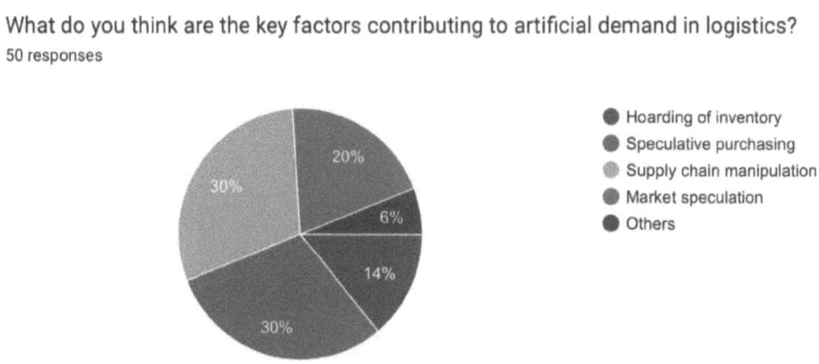

TABLE SHOWING THE EXPERIENCE IN SUPPLY CHAIN DISRUPTION IN ARTIFICIAL DEMAND

TABLE 4.8

S.NO	TIME PERIOD	NO OF RESPONDENTS	PERCENT
1	Frequently	10	20.4%
2	Occasionally	26	53.1%
3	Rarely	9	18.4%
4	Never	4	8.2%
5	Not answered	1	0%
	TOTAL	**50**	**100**

INTERPRETATION:

Primary Data:

The table 4.8 implies that 20.4% of the respondents face it Frequently , 53.1% of the respondents face it Occasionally, 18.4% of respondents face it Rarely and 8.2% of respondents have Never faced it.

Majority of the Respondents (53.1%) have faced it Occasionally.

CHART 4.8

CHART SHOWING THE EXPERIENCE IN SUPPLY CHAIN DISRUPTION IN ARTIFICIAL DEMAND

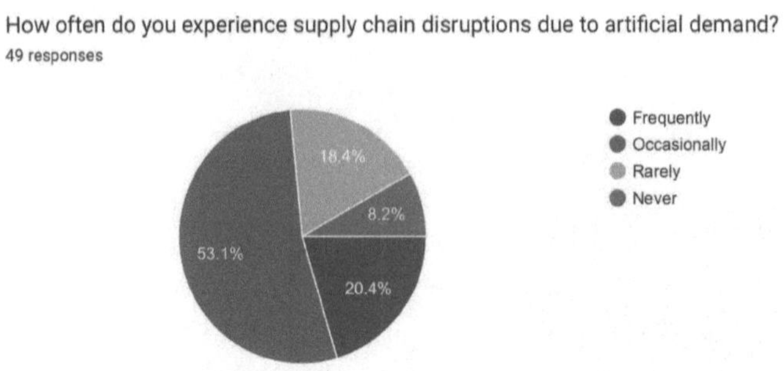

TABLE SHOWING THE PRIMARY IMPACTS OF ARTIFICIAL DEMAND IN LOGISTICS OPERATIONS

TABLE 4.9

S.NO	IMPACTS	NO OF RESPONDENTS	PERCENT
1	Increased costs	6	12%
2	Shortages of goods	23	46%
3	Overstocking issues	10	20%
4	Delayed shipments	7	14%
5	Others	4	8%
	TOTAL	**50**	**100**

INTERPRETATION:

Primary Data:

The table 4.9 implies that 12% of the respondents agree to Increased Costs, 46% of the respondents agree to Shortages of goods, 20% of respondents agree to Overstocking issues, 14% of respondents agree to Delayed shipments and 8% of respondents agree to Other category.

Majority of the Respondents (46%) have agreed to Shortages of goods .

CHART 4.9

CHART SHOWING THE PRIMARY IMPACTS OF ARTIFICIAL DEMAND IN LOGISTICS OPERATIONS

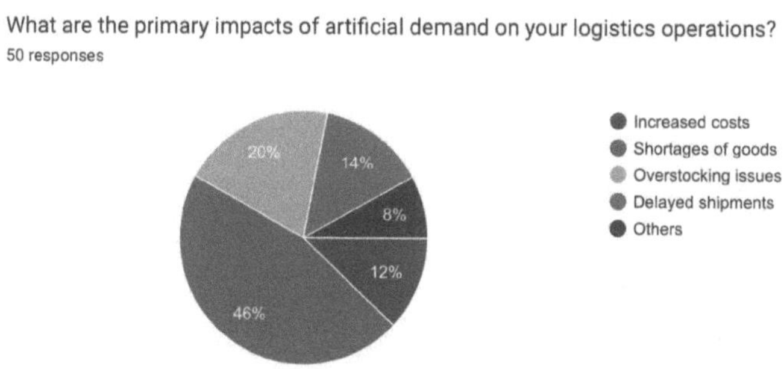

TABLE SHOWING THE AFFECT OF ARTIFICIAL DEMAND IN COMPANY'S PROFITABILITY

TABLE 4.10

S.NO	STATEMENT	NO OF RESPONDENTS	PERCENT
1	Significantly negative impact	10	20.4%
2	Slightly negative impact	21	42.9%
3	No impact	10	20.4%
4	Slightly positive impact	7	14.3%
5	Significantly positive impact	1	2%
6	Not answered	1	0%
	TOTAL	**50**	**100**

INTERPRETATION:

Primary Data:

The table 4.10 implies that 20.4% of the respondents stated Significantly negative impact , 42.9% of the respondents stated Slightly negative impact, 20.4% of respondents stated No impact, 14.3% of respondents stated Slightly positive impact and 2% of respondents stated Significantly positive impact.

Majority of the Respondents (42.9%) have stated Slightly negative impact.

CHART 4.10

CHART SHOWING THE AFFECT OF ARTIFICIAL DEMAND IN COMPANY'S PROFITABILITY

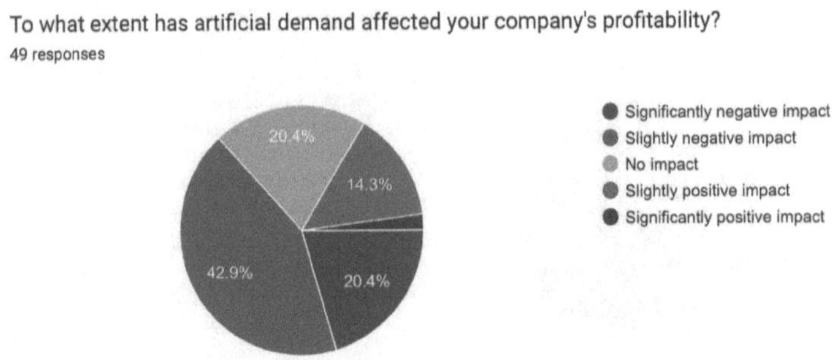

TABLE SHOWING THE KEY DRIVERS OF ARTIFICIAL DEMAND IN LOGISTICS

TABLE 4.11

S.NO	KEY DRIVERS	NO OF RESPONDENTS	PERCENT
1	Consumer panic buying	11	22%
2	Supplier and distributor stockpiling	13	26%
3	Government trade	17	34%
4	Market speculation	3	6%
5	Others	6	12%
	TOTAL	**50**	**100**

INTERPRETATION:

Primary Data:

The table 4.11 implies that 22% of the respondents agree towards Consumer panic buying, 26% of the respondents agree towards Supplier and distributor stockpiling, 34% of respondents agree towards Government trade, 6% of respondents agree towards Market speculation and 12% of respondents have stated Others.

Majority of the Respondents (34%) agree to Government trade.

CHART 4.11

CHART SHOWING THE KEY DRIVERS OF ARTIFICIAL DEMAND IN LOGISTICS

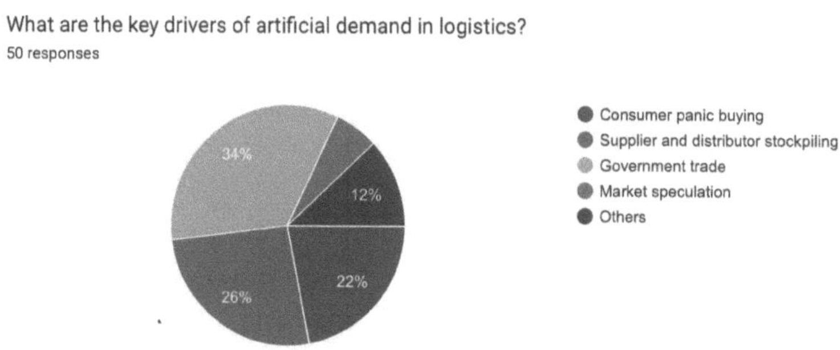

TABLE SHOWING AN INCREASE IN ARTIFICIAL DEMAND DUE TO RECENT GLOBAL TRENDS

TABLE 4.12

S.NO	STATEMENT	NO OF RESPONDENTS	PERCENT
1	Yes	38	79.2%
2	No	10	20.8%
3	Not answered	2	0%
	TOTAL	**50**	**100**

INTERPRETATION:

Primary Data:

The table 4.12 implies that 79.2% of the respondents stated YES and 20.8% of the respondents stated NO.

Majority of the Respondents (79.2%) have stated YES.

CHART 4.12

CHART SHOWING AN INCREASE IN ARTIFICIAL DEMAND DUE TO RECENT GLOBAL TRENDS

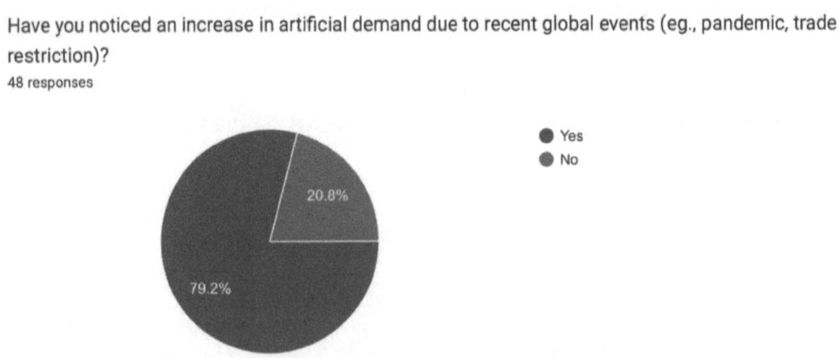

TABLE SHOWING EXTERNAL FACTORS LIKE DIGITAL MARKETING AND MEDIA EXAGGERATE ARTIFICIAL DEMAND

TABLE 4.13

S.NO	STATEMENT	NO OF RESPONDENTS	PERCENT
1	Strongly agree	12	24.5%
2	Agree	13	26.5%
3	Neutral	17	34.7%
4	Disagree	5	10.2%
5	Strongly disagree	2	4.1%
6	Blank	1	0%
	TOTAL	**50**	**100**

INTERPRETATION:

Primary Data:

The table 4.13 implies that 24.5% of the respondents Strongly agree, 26.5% of the respondents Agree, 34.7% of the respondents Neutral, 10.2% of the respondents Disagree and 4.1% of the respondents Strongly disagree.

Majority of the Respondents (34.7%) are Neutral.

CHART 4.13

CHART SHOWING EXTERNAL FACTORS LIKE DIGITAL MARKETING AND MEDIA EXAGGERATE ARTIFICIAL DEMAND

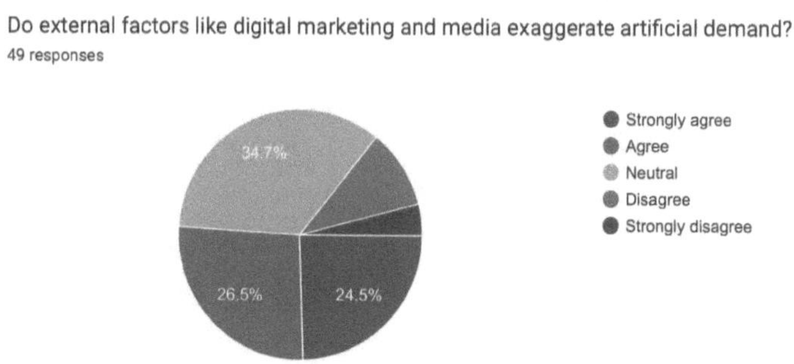

TABLE SHOWING THE COMPANY'S INITIATIVE TO MITIGATE ARTIFICIAL DEMAND

TABLE 4.14

S.NO	MEASURES TAKEN	NO OF RESPONDENTS	PERCENT
1	Demand forecasting and analytics	10	20%
2	Supplier diversification	24	48%
3	Inventory buffer planning	9	18%
4	Supply chain transparency	3	6%
5	Others	4	8%
	TOTAL	**50**	**100**

INTERPRETATION:

Primary Data:

The table 4.14 implies that 20% of the respondents agree to Demand forecasting and analytics, 48% of the respondents agree to Supplier diversification, 18% of the respondents agree to Inventory buffer planning, 6% of the respondents agree to Supply chain transparency and 8% of the respondents agree to Others.

Majority of the Respondents (48%) agree to Supplier diversification.

CHART 4.14

CHART SHOWING THE COMPANY'S INITIATIVE TO MITIGATE ARTIFICIAL DEMAND

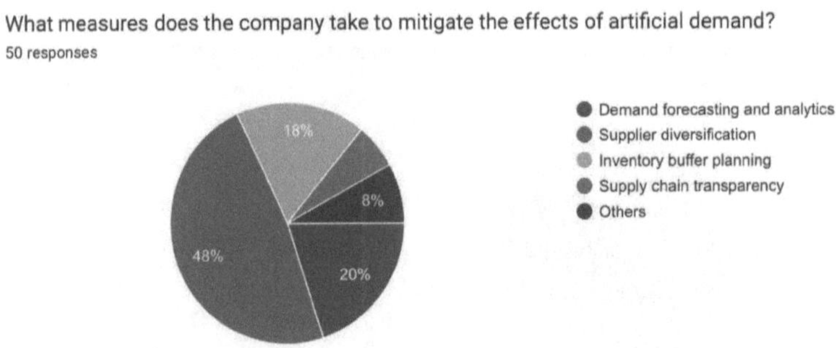

TABLE SHOWING EFFECT OF STARTEGIES IN MANAGING ARTIFICIAL DEMAND

TABLE 4.15

S.NO	STATEMENT	NO OF RESPONDENTS	PERCENT
1	Very effective	17	34.7%
2	Somewhat effective	15	30.6%
3	Neutral	15	30.6%
4	Not effective	2	4.1%
5	Not answered	1	0%
	TOTAL	**50**	**100**

INTERPRETATION:

Primary Data:

The table 4.15 implies that 34.7% of the respondents stated Very effective, 30.6% of the respondents stated Somewhat effective, 30.6% of the respondents stated Neutral and 4.1% of the respondents stated Not effective.

Majority of the Respondents (34.7%) stated Very effective.

CHART 4.15

CHART SHOWING EFFECT OF STARTEGIES IN MANAGING ARTIFICIAL DEMAND

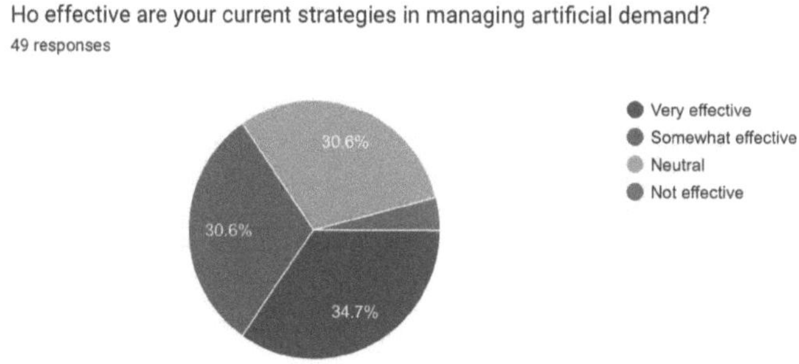

TABLE SHOWING TECHNOLOGY HELP REDUCE ARTIFICIAL DEMAND RELATED TO DISRUPTIONS

TABLE 4.16

S.NO	STATEMENT	NO OF RESPONDENTS	PERCENT
1	Yes, significant	11	22.4%
2	Yes, to some extent	22	44.9%
3	Not sure	13	26.5%
4	No, technology won't help	3	6.1%
5	Not answered	1	0%
	TOTAL	**50**	**100**

INTERPRETATION:

Primary Data:

The table 4.16 implies that 22.4% of the respondents stated Yes, significant, 44.9% of the respondents stated Yes, to some extent, 26.5% of the respondents stated Not sure and 6.1% of the respondents stated No, technology won't help.

Majority of the Respondents (44.9%) have stated Yes, to some extent .

CHART 4.16

CHART SHOWING TECHNOLOGY HELP REDUCE ARTIFICIAL DEMAND RELATED TO DISRUPTIONS

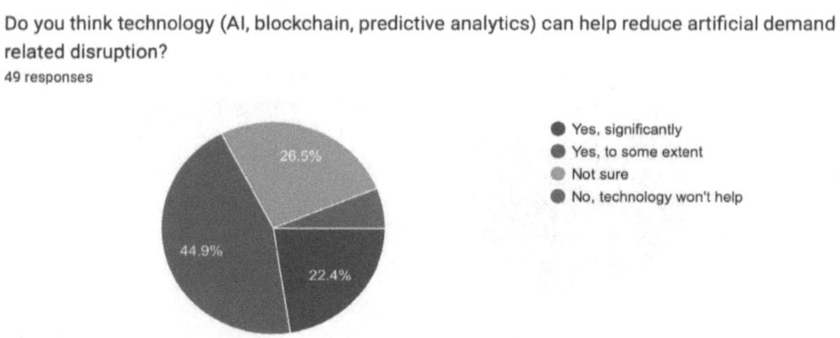

TABLE SHOWING GOVERNAMNT REGULATIONS TO CONTROL ARTIFICIAL DEMAND

TABLE 4.17

S.NO	STATEMENT	NO OF RESPONDENTS	PERCENT
1	Yes, strict regulations are needed	14	28%
2	Yes, but with flexibility	22	44%
3	No, market forces should control it	9	18%
4	Not sure	5	10%
	TOTAL	**50**	**100**

INTERPRETATION:

Primary Data:

The table 4.17 implies that 28% of the respondents have stated Yes, strict regulations are needed, 44% of the respondents stated Yes, but with flexibility, 18% of the respondents stated No, market forces should control it and 10% of the respondents stated Not sure.

Majority of the Respondents (44%) have stated Yes, but with flexibility.

CHART 4.17

CHART SHOWING GOVERNAMNT REGULATIONS TO CONTROL ARTIFICIAL DEMAND

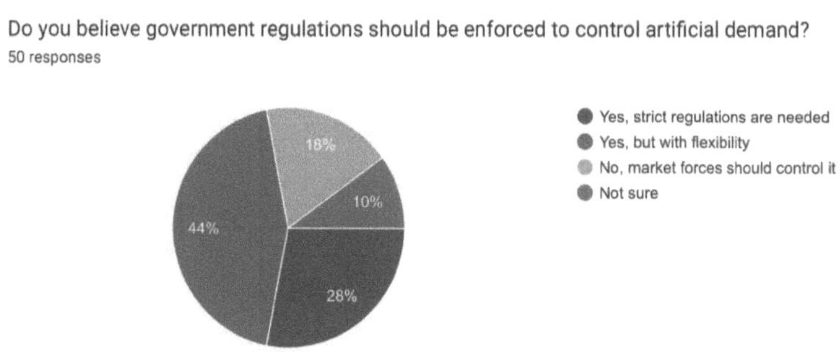

TABLE SHOWING CUSTOMERS PLAY IN CREATING OR MITIGATING ARTIFICIAL DEMAND

TABLE 4.18

S.NO	STATEMENT	NO OF RESPONDENTS	PERCENT
1	A major role	23	46%
2	A minor role	20	40%
3	No role	7	14%
	TOTAL	**50**	**100**

INTERPRETATION:

Primary Data:

The table 4.18 implies that 46% of the respondents agree to A major role, 40% of the respondents agree to A minor role and 14% of the respondents agree to No role.

Majority of the Respondents (46%) have agreed to have A major role.

CHART 4.18

CHART SHOWING CUSTOMERS PLAY IN CREATING OR MITIGATING ARTIFICIAL DEMAND

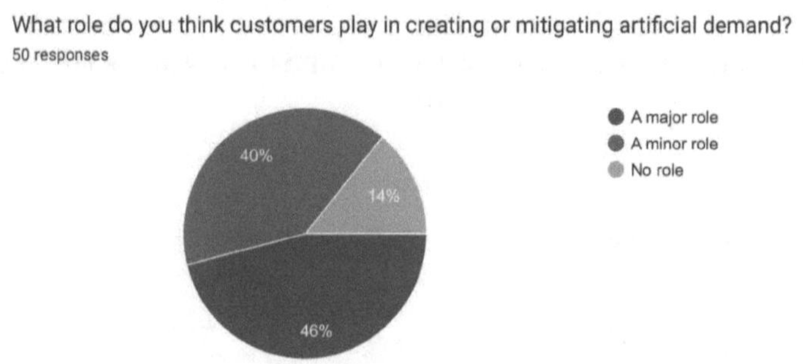

CHAPTER 5
FINDINGS, CONCLUSION AND SUGGESTIONS

Findings

- Majority of the respondents (70%) belong to the age group 18-24.
- Majority of the respondents (71.4%) are Male.
- Majority of the respondents (28.6%) are Supplier.
- Majority of the respondents (42%) have less than 1 year of experience.
- Majority of the respondents (61.2%) are from Medium scale company size.
- Majority of the respondents (44%) have stated Yes.
- Majority of the respondents (30%) have voted from Speculative Purchasing and Supply Chain Manipulation.
- Majority of the respondent (53.1%) have faced it Occasionally
- Majority of the respondents (46%) have agreed to Shortages of goods
- Majority of the respondents (42.9%) have stated Slightly negative impact
- Majority of the respondents (34%) agree to Government trade
- Majority of the respondents (79.2%) have stated Yes.
- Majority of the respondents (34.7%) are Neutral.
- Majority of the respondents (48%) agree to Supplier Diversification.
- Majority of the respondents (34.7%) stated Very effective.
- Majority of the respondents (44.9%) have stated Yes, to some extent.
- Majority of the respondents (44%) have stated Yes, but with flexibility.
- Majority of the respondents (46%) have agreed to have A major role.

Suggestions

- Government Trade should be used towards enhancing the limitations of artificial demand in logistics.
- Supplier diversification and Shortage of goods can be eradicated in order the meet the necessary demand and supply in the field of logistics.

- A specific application can be created for the demand products in the international market to track the artificial products used in logistics.

Conclusions

Implementing a server-based tracking system significantly enhances stock availability management and helps businesses avoid unnecessary purchases of foreign products at inflated prices. Price transparency mechanisms and demand forecasting tools empower consumers to make informed decisions, thereby preventing price exploitation. Establishing long-term relationships with verified third-party agents ensures a stable and reliable supply chain, reducing dependence on unpredictable market conditions. The adoption of AI-driven analytics, blockchain transparency, and regulatory frameworks effectively mitigates artificial demand in logistics, creating a more stable and fair marketplace. By integrating these strategies, businesses and consumers can achieve cost savings, better resource allocation, and enhanced supply chain efficiency, leading to a sustainable and transparent market ecosystem.

BIBLIOGRAPHY

Chopra, S., & Meindl, P. (2019). Supply chain management: Strategy, planning, and operation (7th ed.). Pearson.

Christopher, M. (2016). Logistics & supply chain management (5th ed.). Pearson.

Sheffi, Y. (2015). The power of resilience: How the best companies manage the unexpected. MIT Press.

Journal Articles:

Cachon, G. P., & Lariviere, M. A. (2005). Supply chain coordination with revenue-sharing contracts: Strengths and limitations. Management Science, 51(1), 30-44. https://doi.org/10.1287/mnsc.1040.0215

Ivanov, D. (2021). Supply chain viability and the COVID-19 pandemic: A conceptual and formal generalisation of four major adaptation strategies. International Journal of Production Research, 59(12), 3535-3552. https://doi.org/10.1080/00207543.2021.1875866

Lee, H. L., Padmanabhan, V., & Whang, S. (1997). The bullwhip effect in supply chains. Sloan Management Review, 38(3), 93-102.

Reports and Case Studies:

Gartner. (2022). Artificial demand in retail and manufacturing supply chains. Gartner Research. McKinsey & Company. (2021). The future of supply chain disruptions and demand manipulation. McKinsey Global Institute. https://www.mckinsey.com.

World Economic Forum. (2020). How COVID-19 is reshaping global supply chains. World Economic Forum. https://www.weforum.org

Web Resources:

Harvard Business Review. (2021). Why supply chains create artificial scarcity and demand. Harvard Business Review. https://hbr.org

REFERENCES

- Ali, A., Mahfouz, A., & Arisha, A. (2017). Analysing supply chain resilience: Integrating the constructs in a concept mapping framework via a systematic literature review. *Supply Chain Management: An International Journal, 22*(1), 16–39.

  Article Google Scholar

- Amini, M., & Li, H. (2011). Supply chain configuration for diffusion of new products: An integrated optimization approach. *Omega,39*(3), 313–322.

  Article Google Scholar

- Araz, O., Choi, T., Olson, D., & Salman, F. (2020). Data analytics for operational risk management. *Decision Sciences*. https://doi.org/10.1111/deci.12443.

  Article Google Scholar

- Atadeniz, S. N., & Sridharan, S. V. (2019). Effectiveness of nervousness reduction policies when capacity is constrained. *International Journal of Production Research,58,* 4121.

  Article Google Scholar

- Awasthy, P., Gopakumar, K. V., Gouda, S. K., & Haldar, T. (2019). Trust in humanitarian operations: a content analytic approach for an Indian NGO. *International Journal of Production Research, 57*(9), 2626–2641. https://doi.org/10.1080/00207543.2019.1566652.

  Article Google Scholar

- Baghalian, A., Rezapour, S., & Farahani, R. Z. (2013). Robust supply chain network design with service level against disruptions and demand uncertainties: A real-life case. *European Journal of Operational Research,227*(1), 199–215.

  Article Google Scholar

- Baharmand, H., Comes, T., & Lauras, M. (2019). Defining and measuring the network flexibility of humanitarian supply chains: Insights from the 2015 Nepal earthquake. *Annals of Operations Research, 283*(1), 961–1000.

  Article **Google Scholar**

- Banker, S. (2019). Supply chain trends to watch in 2019. *Forbes, Transportation*https://www.forbes.com/sites/stevebanker/2019/01/05/supply-chain-trends-to-watch-in-2019/#2b4b4f5a323d.

- Banomyong, R., Varadejsatitwong, P., & Oloruntoba, R. (2019). A systematic review of humanitarian operations, humanitarian logistics and humanitarian supply chain performance literature 2005 to 2016. *Annals of Operations Research, 283*(1–2), 71–86.

  Article **Google Scholar**

- Baryannis, G., Dani, S., & Antoniou, G. (2019a). Predicting supply chain risks using machine learning: The trade-off between performance and interpretability. *Future Generation Computer Systems, 101,* 993–1004.

  Article **Google Scholar**

- Baryannis, G., Validi, S., Dani, S., & Antoniou, G. (2019b). Supply chain risk management and artificial intelligence: State of the art and future research directions. *International Journal of Production Research, 57*(7), 2179–2202. https://doi.org/10.1080/00207543.2018.1530476.

  Article **Google Scholar**

- BCI-Business Continuity Institute. (2019). Supply chain resilience 10 year trend analysis. 2009–2018. *Zurich Insurance Group*https://www.b-c-training.com/img/uploads/resources/Supply-Chain-Resilience-10-year-trend-analysis.pdf.

- Ben-Daya, M., Hassini, E., & Bahroun, Z. (2019). Internet of things and supply chain management: A literature review. *International Journal of Production Research,57*(15–16), 4719–4742.

Article Google Scholar

- Bier, T., Lange, A., & Glock, C. H. (2019). Methods for mitigating disruptions in complex supply chain structures: A systematic literature review. *International Journal of Production Research, 58,* 1835.

Article Google Scholar

- Blackhurst, J., Craighead, C. W., Elkins, D., & Handfield, R. B. (2005). An empirically derived agenda of critical research issues for managing supply-chain disruptions. *International Journal of Production Research,43*(19), 4067–4081. https://doi.org/10.1080/00207540500151549.

Article Google Scholar

- Brandon-Jones, E., Squire, B., & Van Rossenberg, Y. G. T. (2015). The impact of supply base complexity on disruptions and performance: The moderating effects of slack and visibility. *International Journal of Production Research, 53*(22), 6903–6918. https://doi.org/10.1080/00207543.2014.986296.

Article Google Scholar

- Braunscheidel, M. J., & Suresh, N. C. (2009). The organizational antecedents of a firm's supply chain agility for risk mitigation and response. *Journal of operations Management,27*(2), 119–140.

Article Google Scholar

- Brintrup, A., Pak, J., Ratiney, D., Pearce, T., Wichmann, P., Woodall, P., et al. (2019). Supply chain data analytics for predicting supplier disruptions: a case study in complex asset manufacturing. *International Journal of Production Research, 58,* 3330.

Article Google Scholar

- Brusset, X., & Teller, C. (2017). Supply chain capabilities, risks, and resilience. International Journal of Production Economics,184, 59–68.

Article Google Scholar

- Cantor, D. E., Blackhurst, J., Pan, M., & Crum, M. (2014). Examining the role of stakeholder pressure and knowledge management on supply chain risk and demand responsiveness. The International Journal of Logistics Management, 25, 202.

  Article **Google Scholar**

- Centobelli, P., Cerchione, R., & Ertz, M. (2019). Managing supply chain resilience to pursue business and environmental strategies. Business Strategy and the Environment, 29, 1215.

  Google Scholar

- Chen, K. B., & Yang, L. (2014). Random yield and coordination mechanisms of a supply chain with emergency backup sourcing. International Journal of Production Research,52(16), 4747–4767. https://doi.org/10.1080/00207543.2014.886790.

  Article **Google Scholar**

- Chen, C. P., & Zhang, C. Y. (2014). Data-intensive applications, challenges, techniques and technologies: A survey on Big Data. Information Sciences, 275, 314–347.

  Article **Google Scholar**

- Chen, H., Chiang, R. H. L., & Storey, V. C. (2012). Business intelligence and analytics: From big data to big impact. MIS Quarterly,36(4), 1165.

  Article **Google Scholar**

- Chongvilaivan, A. (2011). Managing global supply chain disruptions: Experience from Thailand's 2011 flooding. Regional Economic Studies Programme, Institute of Southeast Asian Studies (ISEAS), 30

- Chopra, S., & Sodhi, M. (2004). Supply-chain breakdown. MIT Sloan Management Review,46(1), 53–61.

  Google Scholar

- Chopra, S., & Sodhi, M. (2014). Reducing the risk of supply chain disruptions. MIT Sloan Management Review, 55(3), 72–80.

  Google Scholar

- Chowdhury, M. M. H., & Quaddus, M. (2017). Supply chain resilience: Conceptualization and scale development using dynamic capability theory. International Journal of Production Economics, 188, 185–204.

  Article **Google Scholar**

- Christopher, M., & Peck, H. (2004). Building the resilient supply chain. The International Journal of Logistics Management, 15(2), 1–14.

  Article **Google Scholar**

- Craighead, C. W., Blackhurst, J., Rungtusanatham, M. J., & Handfield, R. B. (2007). The severity of supply chain disruptions: Design characteristics and mitigation capabilities. Decision Sciences,38(1), 131–156.

QUESTIONNAIRES

1. What is your role in the supply chain/logistics industry?
   - a) Manufacturer
   - b) Supplier
   - c) Distributor
   - d) Retailer
   - e) Other (please specify)

2. How many years of experience do you have in logistics/supply chain management?
   - a) Less than 1 year
   - b) 1-3 years
   - c) 4-7 years
   - d) 8+ years

3. What is the size of your company's supply chain operations?
   - a) Small-scale
   - b) Medium-scale
   - c) Large-scale

4. Have you encountered artificial demand in your supply chain operations?
   - a) Yes
   - b) No
   - c) Not sure

5. What do you think are the key factors contributing to artificial demand in logistics?
   - a) Hoarding of inventory
   - b) Speculative purchasing
   - c) Supply chain manipulation
   - d) Market speculation
   - e) Other (please specify)

6. How often do you experience supply chain disruptions due to artificial demand?
   - a) Frequently
   - b) Occasionally
   - c) Rarely
   - d) Never

7. What are the primary impacts of artificial demand on your logistics operations?
   - a) Increased costs
   - b) Shortages of goods
   - c) Overstocking issues

- d) Delayed shipments
- e) Other (please specify)

8. To what extent has artificial demand affected your company's profitability?

- a) Significantly negative impact
- b) Slightly negative impact
- c) No impact
- d) Slightly positive impact
- e) Significantly positive impact

9. What are the key drivers of artificial demand in logistics?

- a) Consumer panic buying
- b) Supplier and distributor stockpiling
- c) Government trade policies
- d) Market speculation
- e) Other (please specify)

10. Have you noticed an increase in artificial demand due to recent global events (e.g., pandemic, trade restrictions)?

- a) Yes
- b) No

11. Do external factors like digital marketing and media exaggerate artificial demand?

- a) Strongly agree
- b) Agree
- c) Neutral
- d) Disagree
- e) Strongly disagree

12. What measures does your company take to mitigate the effects of artificial demand?

- a) Demand forecasting and analytics
- b) Supplier diversification
- c) Inventory buffer planning
- d) Supply chain transparency
- e) Other (please specify)

13. How effective are your current strategies in managing artificial demand?

- a) Very effective
- b) Somewhat effective
- c) Neutral

- d) Not effective

14. Do you think technology (AI, blockchain, predictive analytics) can help reduce artificial demand-related disruptions?

- a) Yes, significantly
- b) Yes, to some extent
- c) Not sure
- d) No, technology won't help

15. Do you believe government regulations should be enforced to control artificial demand?

- a) Yes, strict regulations are needed
- b) Yes, but with flexibility
- c) No, market forces should control it
- d) Not sure

16. What role do you think customers play in creating or mitigating artificial demand?

- a) A major role
- b) A minor role
- c) No role

17. Suggestion _______________

FSC
www.fsc.org
MIX
Papier | Fördert gute Waldnutzung
FSC® C083411

Kolibri 360 GmbH
Ferdinand-Jühlke-Straße 7
99095 Erfurt, Deutschland
produktsicherheit@kolibri360.de

Druck:
CPI Druckdienstleistungen GmbH
im Auftrag der
Kolibri 360 GmbH
Ferdinand-Jühlke-Str. 7
99095 Erfurt